A PRACTICAL SOLUTION TO RUBIK'S MAGIC™

by

ASHWIN BELUR and BLAIR WHITAKER

A DELL BOOK

CONTENTS

To our parents:

For posing the problems and puzzles of life in a form easy enough to understand, yet tough enough to challenge

---

Published by
Dell Publishing Co., Inc.
1 Dag Hammarskjold Plaza
New York, New York 10017

ISBN: 0-440-17531-3

Printed in the United States of America

November 1986

10 9 8 7 6 5 4 3 2 1

WFH

1 SPECIAL INTRODUCTION

This book contains a solution to my new Rubik's Magic™ puzzle. Like my famous Cube, my new puzzle has a large number of possible solutions, some of them more involved and intricate than others, and a great many play possibilities.

Although the solution I personally favor differs from the one presented here, I commend the authors for the ingenuity they have shown. I was especially impressed with the speed with which they solved the puzzle and the clarity with which they set out their solution. Reading this book will help you to understand some of the intricacies of my puzzle. After you have finished it, you may well want to work out your own solution—or solutions.

Ernö Rubik

AUTHOR'S INTRODUCTION

If you have been trying to solve Rubik's Magic,™ you probably have found that it is not as easy as it looks. After playing with the puzzle for a while, you may have tangled it into a web or gone through a series of colorful, but disconnected, patterns.

In the process of working with the puzzle we have found that there are many paths to the solution. Some are shorter, while others are longer. We have chosen to present a solution which allows you to wander in the space of Magical shapes, while enjoying the color and geometry of the puzzle. We challenge the reader to explore the possibilities.

This book will provide you with an easy-to-follow solution to Rubik's Magic that really works and can be easily understood.

There are step-by-step directions that will help you to Link the Rings,™ accompanied by a pictorial description of the sequences of moves required. The sequence of moves that Link the Rings is short and easy to understand, and with practice you will be able to reach the solution in only a matter of seconds.

2 RUBIK'S PUZZLES

Professor Ernö Rubik of the Academy of Design and Crafts in Budapest, Hungary, invented Rubik's Cube™, which went on to become a worldwide phenomenon. Cubes sold everywhere in the world—the United States, Great Britain, China, and the Union of the Soviet Socialist Republics. The cube's popularity was based on its geometry and color rather than language or style. Because it required no language, interest in it crossed all boundaries. In America kids of all ages were challenged to restore the cube. Contests appeared in shopping malls all over, with victory going to those who could solve the cube the quickest. Rubik, a professor of architecture and design, designed a masterpiece in Rubik's Cube. It provided motion that seemed impossible as well as a puzzle that was colorful and could be moved into an enormous number of combinations.

Now, six years later, Ernö Rubik is back with Rubik's Magic. When removed from its package, it is a rectangular panel of eight tiles arranged into two rows of four each. As with the cube, this puzzle is exciting because it crosses all borders and lives in the world of geometry, perception and pattern. It is also unique in its mechanical properties. The hinges that are used to hold it together, or actually that allow it to move so freely, are interesting in themselves.

HINGE MECHANICS

Folding two tiles together until their faces touch, and moving them apart again, can move a hinge joint from one edge to

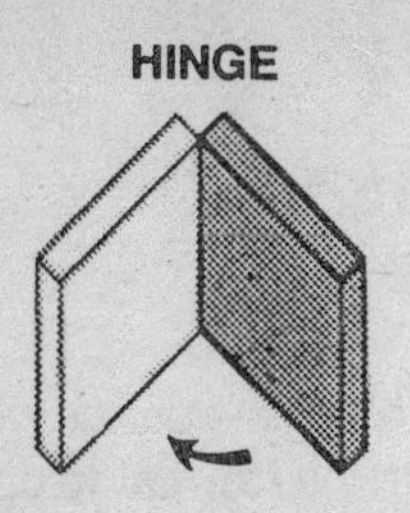

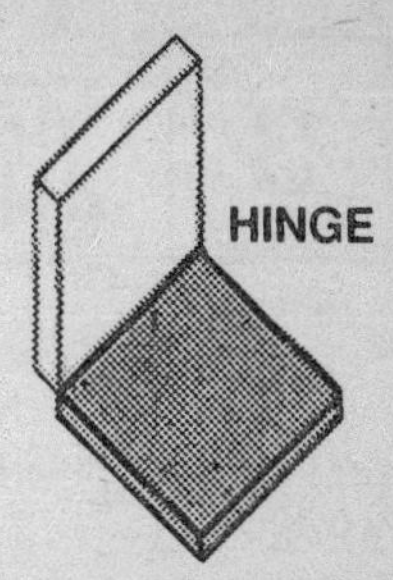

another. As shown in the figure **1** above, the two tiles are folded until their faces meet. In figure **2** the shaded tile is flipped downward. Note that from figure **1** to figure **3** the hinge has actually moved its position. This hinge motion is what enables the puzzle to change its shape and geometry. Detailed examples of this will be seen throughout the book.

If you have Magic in a **rectangular state** of two rows of four tiles each, you will see that one side of Magic has three tiles with intersecting arcs, and the other side has no tiles with intersecting arcs. The side *without* any intersecting arcs will be called the **front side**. The side *with* intersecting arcs will be called the **back side**.

Through a sequence of moves to be described further on, a **rectangular state** can be manipulated into the **rectangular solution** (see photo 2). Magic can also be changed through a longer sequence of moves into the **square solution** (see photo 3) with one row of two tiles and two rows of three tiles. Included is a color insert with pictures of both the **rectangular solution** and the **square solution**.

All the moves required to move from any **rectangular state** to the **rectangular solution** and from the **rectangular solution** to the **square solution** will be described in the next sections. If your puzzle is in a confused state, do not worry, read on and do as much as you can. You will have a chance to read through the next section again with your puzzle in a **rectangular state**.

3 NOTATION AND TERMINOLOGY

A standardized convention for approaching the puzzle will now be described. Place it on a table with the **front side** up. You should see two rows of four tiles each. We will refer to the tiles by numbers. Look at the color pictures in the color insert; they show both the **rectangular solution** and the **square solution**. Notice that each tile has a number associated with it. The numbers on the **front side** of the puzzle are in regular text **(1 2 3 4 5 6 7 8)** . The numbers on the **back side** of the puzzle are in a special text. (1 2 3 4 5 6 7 8).When looking at the diagrams further on in this book you can use the numbers shown to determine which tiles should be seen where as well as their orientation. A number can be oriented in one of four position: up (**4**), right (**4**), down (**4**) and left (**4**). The orientation of a number reflects the actual orientation of the tile it is associated with. So, if a number on a diagram is shown rotated, then the associated tile will be seen with the same rotation.

Look at your puzzle and locate the easily recognizable tiles described below. On the **front side** there are three tiles that have distinguishing features. These features will allow you to identify these three tiles easily. Front tile number **5** has Rubik's signature on it. Front tile number **6** has "PAT. PEND. © 1986 RUBIK MADE IN CHINA" written on it. Front tile number **7** has "MATCHBOX® " written on it. On the **back side** four tiles have distinguishing features. Back tile number **1** has "MATCH-BOX® " and a pair of intersecting arcs printed on it. Back tile number **2** has Rubik's signature written on it. Back tile number **3** has "PAT. PEND. © 1986 RUBIK MADE IN CHINA" and a pair of intersecting arcs. Back tile number **6** is the only tile with three arcs and two intersections. Before you continue be sure you have identified all of the tiles described above.

To aid your recognition of tiles and their numbers, sticky labels may be applied to the tiles, using the numbering scheme shown in the color pictures. This system is actually a great help in first memorizing the sequence of moves. The labels should be small enough so as not to cover the printed colored rings or the grooves on the tiles. Use two different styles of text to distinguish the **front side** from the **back side** of the tiles. Be careful to use the correct orientation as shown in the color pictures. The labels will help you while you are learning the moves but can be removed as you become more familiar with the puzzle.

After locating the tiles described above, the other tiles can also be found easily. Two tiles joined by a hinge will be referred to as neighbors. In the figure below, neighbors are connected by a line. The figure shows all neighbor connections.

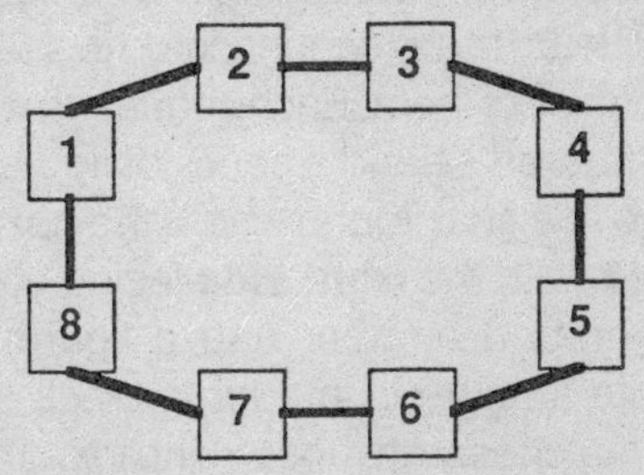

As an example, notice that tile number **7** always has tiles **6** and **8** as neighbors. Also notice that because of the numbering scheme devised and shown in the color pictures, tile number **7** on the **front side** is also tile number **7** on the **back side**.

Before presenting the solution techniques, we caution you not to twist or force tiles in a direction that creates resistance. If a tile does not move freely, try another move.

4 INTRODUCTION TO MOTION

The following sets of instructions are given in an order that will teach you to complete the rings on the **front side** and Link the Rings on the **back side** of the puzzle. It is set up in a way that teaches you some simpler moves first. This allows you to slowly learn the notation and gain a feel for how Magic moves.

These instructions are to be used in parallel with the figures that show the puzzle. As it is changed from one orientation to the next, be aware that the numbers used to describe the orientation should correspond with the tiles you see on your puzzle. The proper numbers must not only be in the proper position relative to the other numbers, but they must also be in the proper orientation, as shown in the figures.

A sequence of moves is made up of a number of individual moves. You will learn many sequences in this section. Be sure you can perform all of them. Each sequence will have a starting figure numbered **1** and a series of figures leading you move by move to a final figure, where the puzzle is in a **rectangular state**, with the **front side** facing you.

Each move has a number and a figure. The number associated with a figure is in a large black square. These numbers are matched with those in the text. The text will use descriptive words such as flip, fold, spin, unfold, lift, and open. The figures will also have arrows showing what tiles to move and which direction to move them in.

Make each move slowly and do not make any extra moves. Read the instructions carefully and do not skip any sections unless you are instructed to do so. Now continue on and always be careful that your puzzle looks like that shown in the figures.

This section is for those who have the puzzle in the **rectangular solution** with the rings complete but not linked. If your puzzle is in a **rectangular state** but not in the **rectangular solution**, then you should turn to the section titled "**Motion of Tiles in a Rectangular State.**" If your puzzle is not in a **rectangular state**, then see the section titled "**Help for Those Who Get Lost**." If at any time you cannot match your puzzle to the next figure, then stop and slowly try to achieve a match; your puzzle must match that figure before you continue.

This first set of moves is easy to learn, yet is important in completing the rings on the **front side** and Linking the Rings on the **back side**. It is important that before starting this set of moves the puzzle is in the starting position shown in figure **1**. Do this entire sequence once and be sure that the puzzle is in the exact position shown.

Before you begin, note that numbers on the **front side** of the puzzle are in solid black, while those on the other side are drawn in a special text.

FIBO

Front **I**n **B**ack **O**ut

1 With Magic as shown, the right and left two tiles are folded in toward the center by pushing them slowly from behind.

2 Continue folding the tiles in toward the center until the puzzle looks like the figure. Check the numbers on your puzzle to be sure they are the same as those shown.

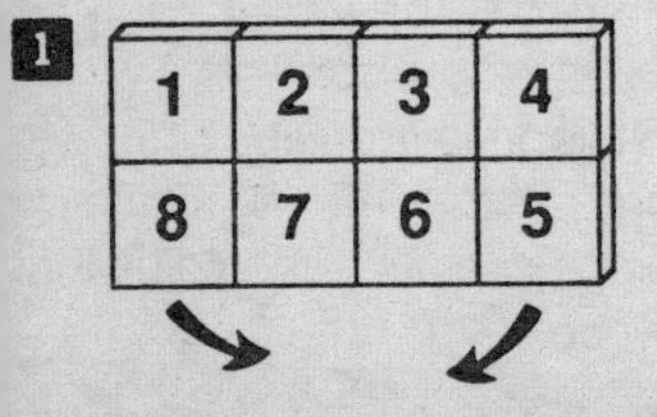

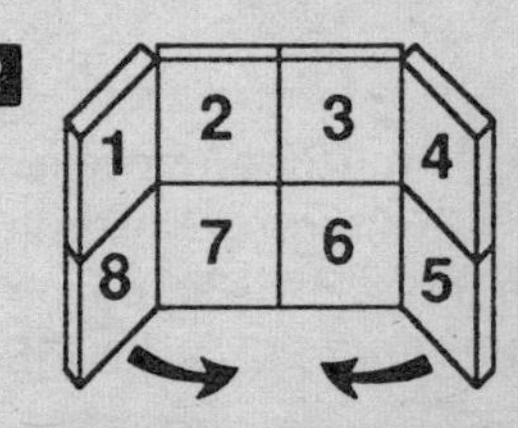

3 Lift from the back the top two tiles; fold down the back bottom two tiles; notice how the hinges have changed from their original positions.

4 Continue opening the top and bottom until it reaches the **rectangular state**.

5 Flip Magic over and spin it until it matches the state and orientation shown.

6 You have now completed your first sequence of moves; this sequence will be referred to as **FIBO** in the rest of this book.

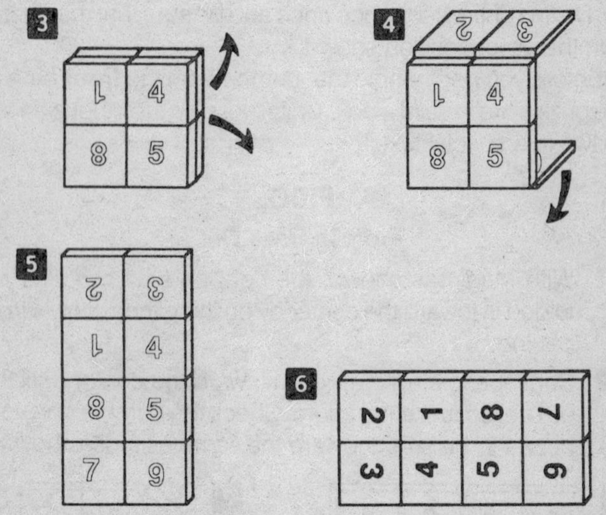

*To continue from here, start with this orientation and complete **FIBO** again. Notice that the numbers this time will not match those in the figure, but the ending orientation should be that of figure 1 of the **FIBO** sequence.

EVEN RING 2

1 With Magic as shown, fold the top four tiles forward and down.

2 Continue folding the top four tiles down until the state shown is reached. Check the numbers to be sure these moves have been done correctly.

3 This new configuration is opened up into a ring shape by pulling back the tiles from the back center hinge and by pulling forward on the front center hinge, as shown by the arrows in the figure.

4 Continue pulling back and forward, but now also push in on the sides until the new state is obtained.

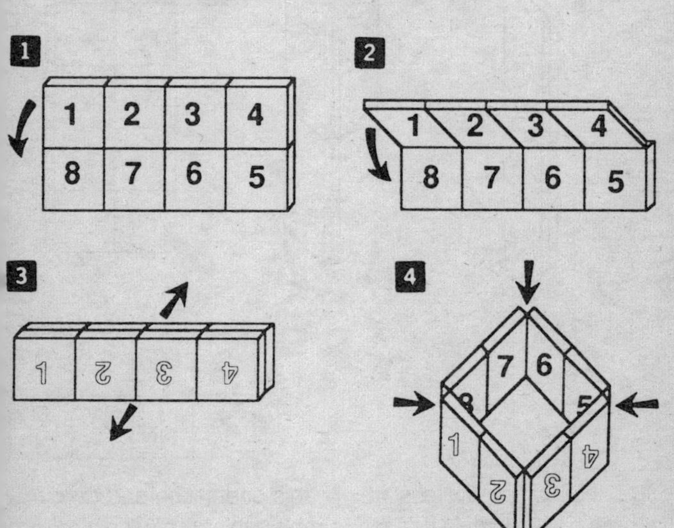

5 Unfold this new state by simultaneously lifting the left four tiles and the right four tiles. You should be checking the numbers at each step to be sure you are making the proper moves.

6 Continue opening until the **rectangular state** is reached once again. Notice that all numbers that are showing are back numbers.

7 Flip the puzzle over and spin it until it matches the state and orientation shown.

8 This is the completion of the **even ring 2** sequence.

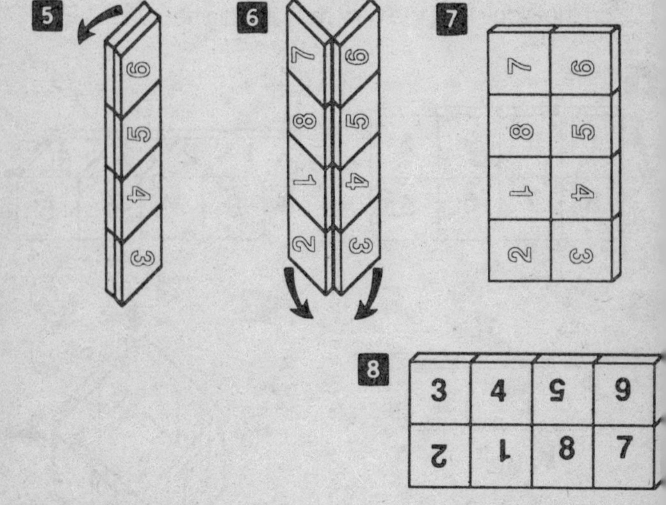

To continue from here, start with this orientation and do **even ring 2** again. Notice that the numbers this time will not match, but the ending orientation will be the same as that in figure **1** of the **even ring 2** sequence.

EVEN RING 1

1 With Magic as shown, slowly fold the top four tiles forward and down.

2 Continue folding the top four tiles down until you achieve the state shown. Check the numbers to be sure these moves have been done correctly.

3 This new configuration is opened up into a ring shape by pulling back on the first hinge on the back side of the folded puzzle while simultaneously pulling forward on the hinge which is third from the left on the front, as the arrows show in the figure.

4 Continue pulling back and forward, but now also push in on the sides until the new state as shown in the next figure is obtained.

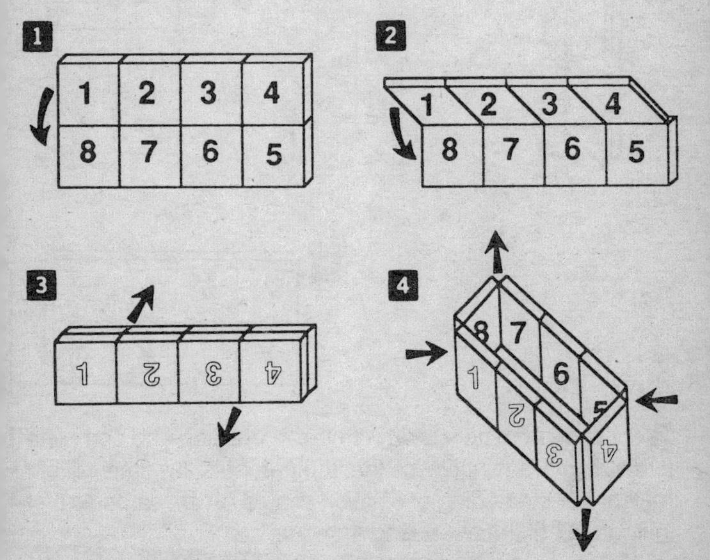

5 Open to this new state by folding left the left four tiles and right the right four tiles. You should be checking the numbers at each step to be sure you are making the proper moves.

6 Continue opening until the **rectangular state** is reached once again. Notice that all numbers showing are front numbers.

7 Spin Magic until it matches the state and orientation shown.

8 This is the completion of the **even ring 1** sequence.

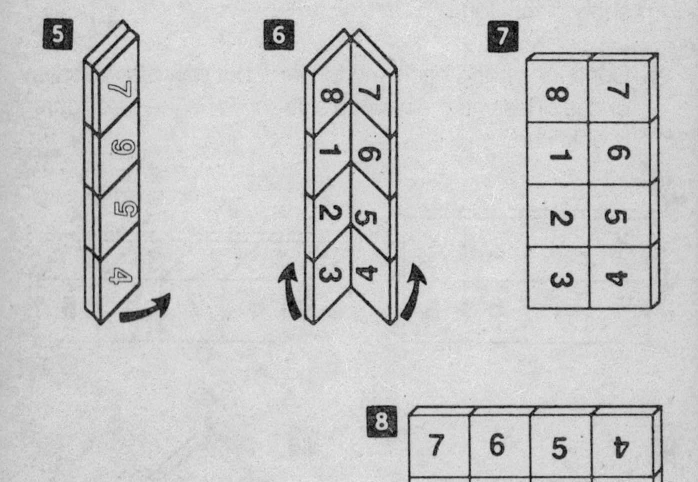

To continue from here, start with this orientation and complete **even ring 1** again. Notice that the numbers this time will not match, but the ending orientation should be the same as that in figure **1** of the **even ring 1** sequence.

EVEN RING 3

1 With Magic as shown, slowly fold the top four tiles forward and down.

2 Continue folding the top four tiles down until you achieve the state shown. Check the numbers to be sure these moves have been done correctly.

3 This new configuration is opened up into a ring shape by pulling back on the third hinge from the left on the back side of the folded puzzle while simultaneously pulling forward on the first hinge on the left front side, as the arrows show in the figure.

4 Continue pulling back and forward, but now also push in on the sides until the new state as show in the next figure is obtained.

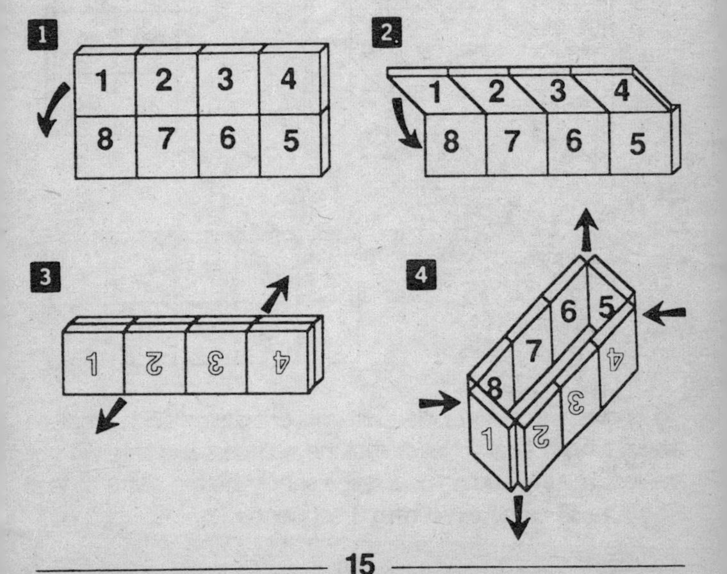

5 Open to this new state by folding left the left four tiles and right the right four tiles. You should be checking the numbers at each step to be sure you are making the proper moves.

6 Continue opening until the **rectangular state** is reached once again. Notice that all numbers that are showing are front numbers.

7 Spin Magic until it matches the state and orientation shown.

8 This is the completion of the **even ring 3** sequence.

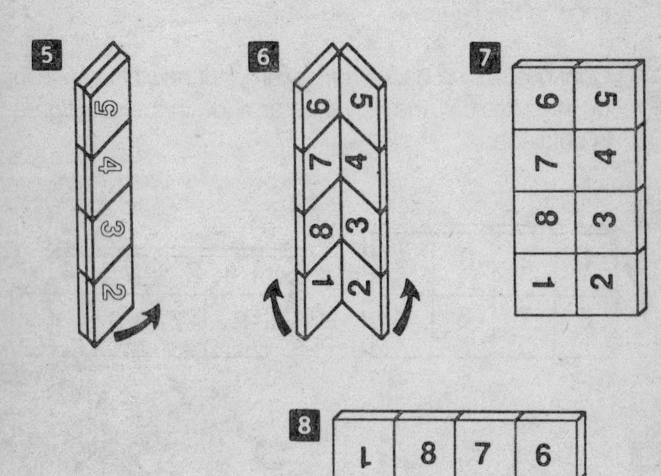

To continue from here, start with this orientation and complete **even ring 3** again. Notice that the numbers this time will not match, but the ending orientation should be the same as that in figure **1** of the **even ring 3** sequence.

1a

Configurations you will encounter as you solve the puzzle.

1a. FISH

1b. BOX WITH LID

1c. OPENING OF "L"

1b

1c

2. RECTANGULAR SOLUTION
(solution of front side)

3. SQUARE SOLUTION
(solution of back side)

Any of these configurations mean that you are not on the proper path to the solution.

4a. RING

4b. HAT

4c. BUTTERFLY

BIFOR

Back In Fold Open Ring

1. With the puzzle as shown, slowly fold the right and left two tiles backward toward the center.

2. Continue folding the tiles in toward the center until the puzzle looks like that shown in the figure. Check the numbers on your puzzle to be sure they are the same as those shown.

3. Next fold the top four tiles back and down.

4. Continue folding the top four tiles back and down until the puzzle looks like that of the next figure.

5. This next move requires that you pull forward on the front two tiles while simultaneously pulling back on the back two tiles. This should open the puzzle up into a ring shaped like the one shown in the next figure.

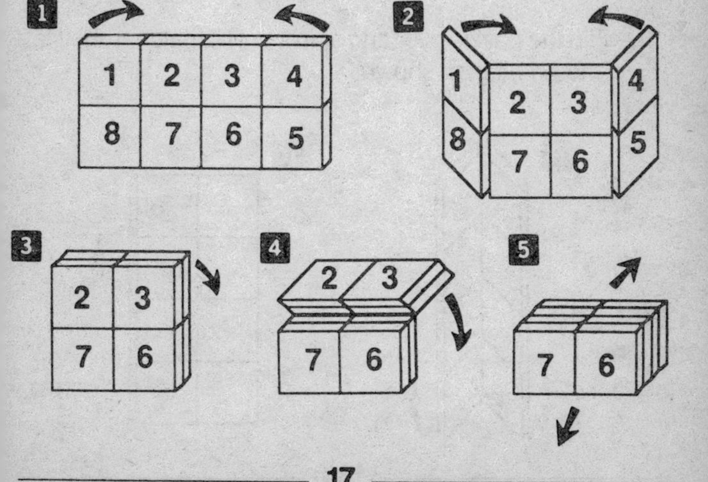

6 Continue opening by pulling forward and back until the square ring is reached.

7 Simultaneously pull on the front center hinge and the back center hinge until the next figure is reached.

8 Open the puzzle by unfolding the left and right four tiles.

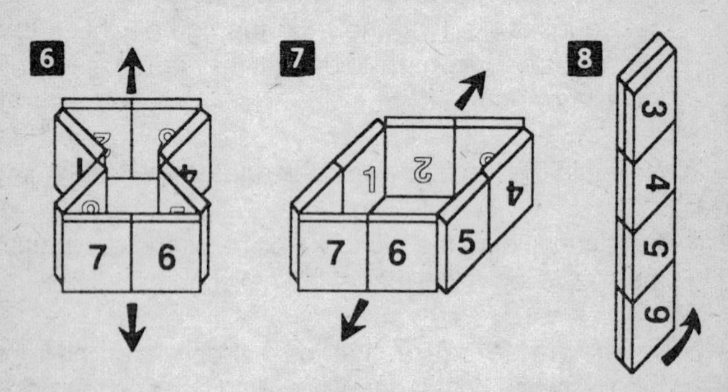

9 Continue unfolding until the puzzle is rectangular.

10 Flip the puzzle over and spin it until it matches the state and orientation shown.

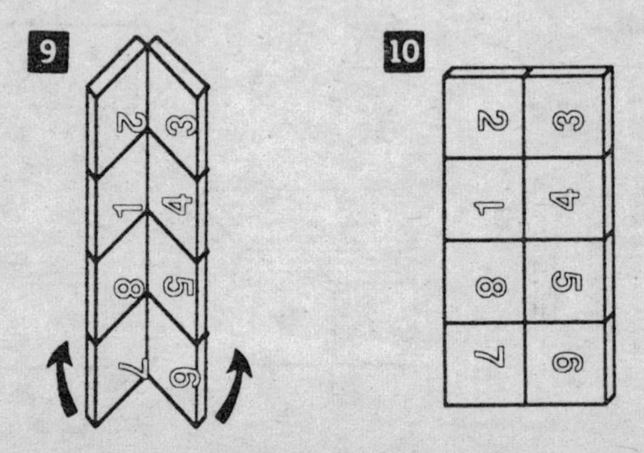

11 This is the most complicated sequence of moves shown yet and is called **BIFOR**. It requires a good understanding of the puzzle's motion for you to have achieved this. Continue reading, but leave the puzzle in its current state until you are instructed to move it again.

The **rectangular state** of Rubik's Magic is very important to understand. It will allow you to practice the moves you have learned so far. But first you must learn a few more moves that are very similar to the ones you have already mastered. The chapter titled "**Motion in a Rectangular State**" provides instructions which use the names of the sequences you have learned thus far. Without changing the state of the puzzle, continue on and learn the remaining sequences of moves.

ODD RING 2

1 With Magic as shown, fold the top four tiles backward and down.

2 Continue folding the top four tiles down until the state shown is reached. Check the numbers to be sure these moves have been done correctly.

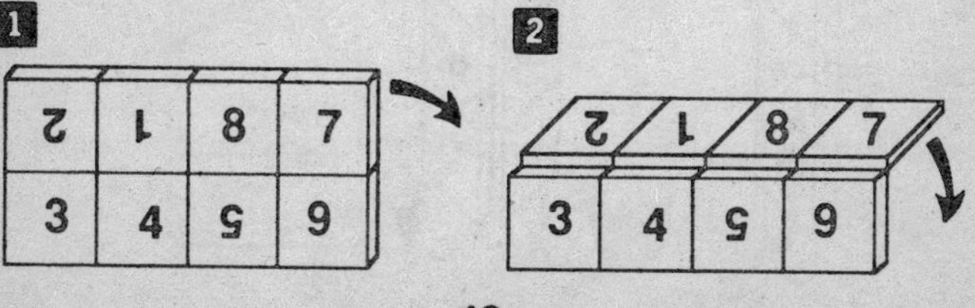

3 This new configuration is opened up into a ring shape by pulling back on the back center hinge and forward on the front center hinge, as shown by the arrows in the figure.

4 Continue pulling back and forward, but now also push in on the sides until the new state is obtained.

5 Unfold this new state by opening left the left four tiles and right the right four tiles. You should be checking the numbers at each step to be sure you are making the proper moves.

6 Continue opening until the rectangular state is reached once again. Notice that all numbers that are showing are front numbers.

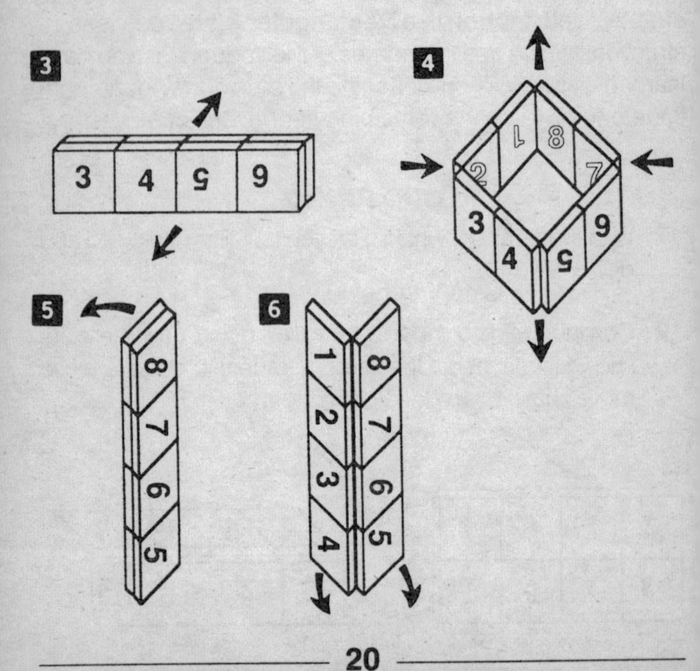

7 Spin Magic until it matches the state and orientation shown.

8 This is the completion of the **odd ring 2** sequence.

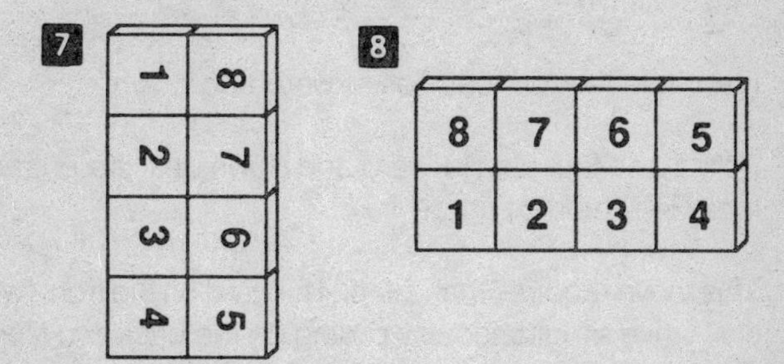

To continue from here, start with this orientation and complete **odd ring 2** again. Notice that the numbers this time will not match, but the ending orientation should be the same as that in figure **1** of this sequence.

FIFOR

Front **I**n **F**old **O**pen **R**ing

1 With the puzzle as shown, slowly fold the right and left two tiles forward toward the center.

2 Continue folding the tiles in toward the center until the puzzle looks like that shown in the next figure. Check the numbers on your puzzle to be sure they are the same as those shown.

3 Next, fold the top four tiles forward and down.

4 Fold these four tiles forward and down until the puzzle looks like the next figure.

5 This move requires that you pull forward on the front two tiles while simultaneously pulling on the back two tiles. This should open the puzzle up into a ring shaped as shown in the next figure.

6 Continue opening by pulling forward and back until the square ring is reached.

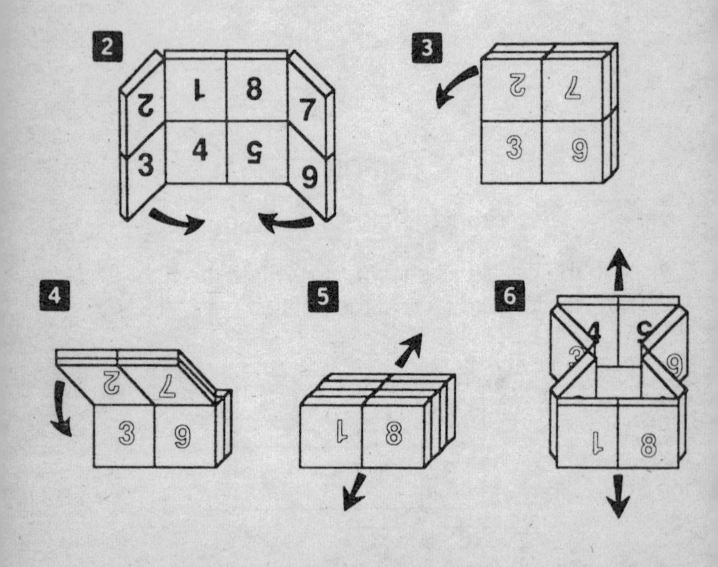

7 Simultaneously pull on the front center hinge and the back center hinge until the next figure is reached.

8 Open the puzzle by unfolding the left and right four tiles.

9 Unfold until the puzzle is in a **rectangular state**.

10 Flip the puzzle over, and spin it until it matches the state and orientation shown.

11 This sequence is called **FIFOR**.

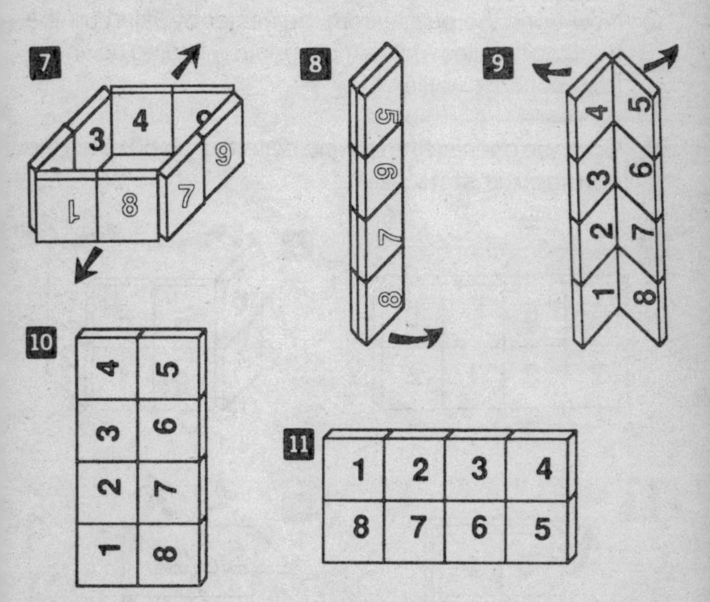

There is only one more sequence to learn, but before you learn it you must change the state of the puzzle. The rings on the front side should now be complete. From this state, you must now repeat the **BIFOR** sequence once before you go on.

BIFO

Back **I**n **F**ront **O**ut

1 With the puzzle as shown in the figure, slowly fold backward the right and left two tiles.

2 Continue folding the tiles backward toward the center until the puzzle looks like that in the next figure shown. Check the numbers on your puzzle to be sure they are the same as those shown.

3 Now open the puzzle from the center by lifting up the front top two tiles while simultaneously folding down the front bottom two tiles.

4 Continue opening the top and bottom until you reach the **rectangular state**.

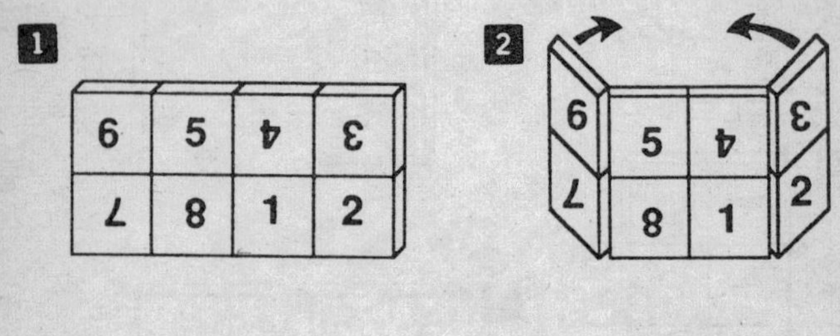

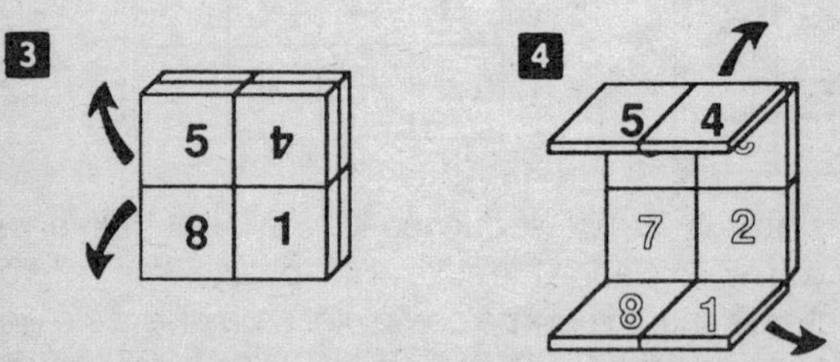

5 Flip the puzzle over and spin it until it matches the state and orientation shown.

6 This sequence will be referred to as **BIFO** in the rest of this book.

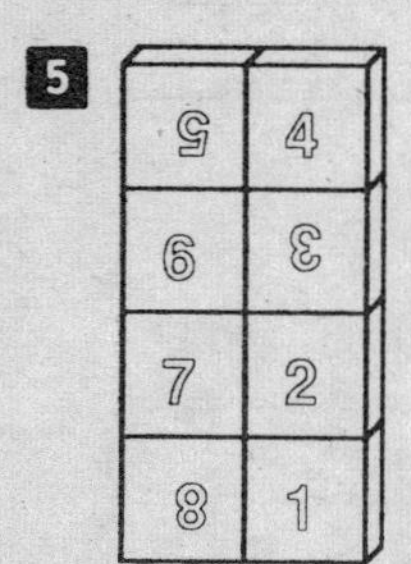

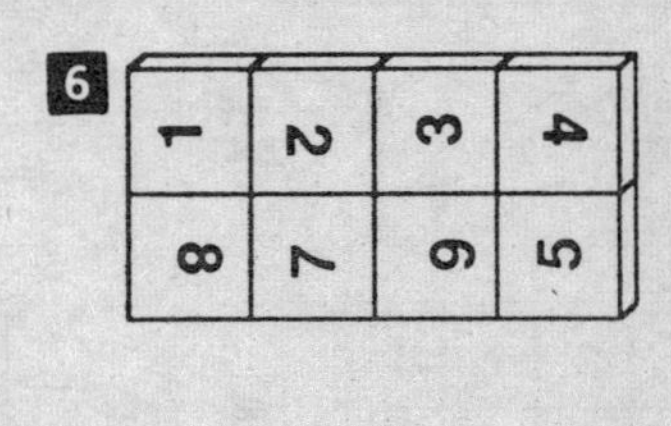

To continue from here, start with the orientation exactly as it appears in figure 6 and repeat **BIFO**. Notice that the numbers this time will not match those shown in the figures, but the ending orientation will be that of figure 1. This puts you back into the position to start the sequence **FIFOR**. Repeat **FIFOR** and you will have returned to the **rectangular solution**.

5 MOTION OF TILES IN A RECTANGULAR STATE

This section describes the motion of the puzzle in a **rectangular state**. If your puzzle is in the **rectangular solution** and you have learned all the sequences described in the previous section, turn to the charts titled "**Motion of Tiles in a Rectangular State**." Explore the possibilities of the **rectangular state**, repeating the sequences previously learned and following the instructions as directed in the chart by the arrows.

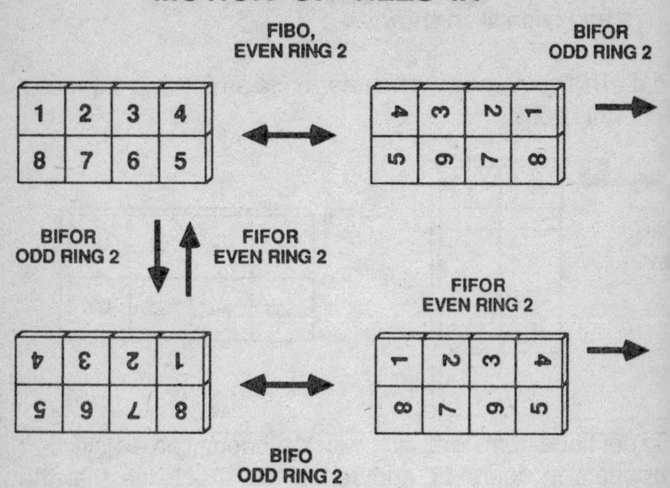

If your puzzle is in a **rectangular state** but not in the **rectangular solution**, then fold it lengthwise. Try to open it into a circular ringlike shape. (If it does not open, then unfold it and fold it in the opposite direction. Now open it into a circular shape.) Your puzzle should now look like photo **4a**. Once you have reached the ringlike state, look at the tiles numbered **5**, **6**, **7**, and **8**. Fold the ring so that the tiles numbered **5**, **6**, **7**, and **8** are all in a row on the same side. Now unfold the puzzle into a **rectangular state**. Your puzzle will now be in one of the configurations shown on the chart.

In finding your puzzle's configuration on the chart, notice that every diagram on the chart shows the puzzle with the **front side** (the side with unlinked rings) up and the top row containing the tiles numbered **1**, **2**, **3**, and **4**. (Notice that the top left diagram is the **rectangular solution**.) After finding your position on the chart, use the sequences as indicated by the arrows to lead you to the **rectangular solution**.

A RECTANGULAR STATE

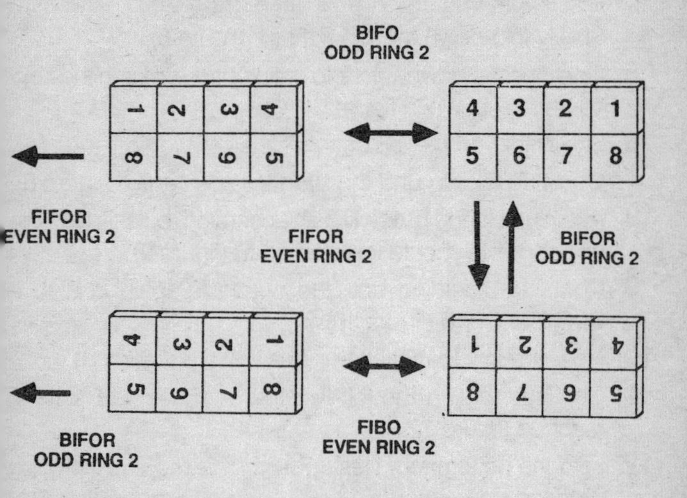

6 THE SOLUTION TO RUBIK'S MAGIC

After learning the sequences of moves that have been introduced in the previous section, you can proceed to the solution of Rubik's Magic on the **back side** and Link the Rings in the **square state**. Instructions leading you to the **square solution** require that you understand the sequences that have already been introduced. The solution as presented assumes that you start with the puzzle in the **rectangular solution**, so you must return to that state before proceeding further.

1. Starting with the puzzle in the **rectangular solution** as shown, perform the **FIBO** sequence. This will put you into the next state.
2. Fold the puzzle like an accordion so that the top profile looks like a **W** (figure 3).

3 Continue folding until the puzzle appears to be a **block** of tiles (figure 4).

4 Slowly fold down the top right tile (figure 5).

5 Continue the downward fold until your puzzle is in the configuration shown in figure 6.

6 Slowly fold up the left bottom tile as shown in figure 7.

7 Continue folding until the **block** is seen again (figure 8).

8 Again unfold the **block** like an accordion, but this time the upper profile should look like an **M** (figure 9).

9 Continue unfolding until the puzzle is again in a **rectangular state** (figure 10).

10 Now slowly fold the left two tiles forward (figure 11).

11 Continue folding the left two tiles until the next state is reached (figure 12).

12 Fold the bottom four tiles upward. This move is tricky, so proceed slowly (figure 13).

13 Continue folding the bottom four tiles up until you achieve the position shown in figure 14.

14 Open up the left side of the puzzle by pulling leftward from the middle.

15 Separate and open up the right two tiles while simultaneously pulling lightly on the left center hinge. The right two tiles become the tail of the next position, the **FISH** (figure 16). (See photo 1a)

16 Continue opening up the two tiles of the **FISH** tail and push lightly on the left center hinge until the next state is reached (figure 17). This is called the **BOX**. As always, be careful to check that the proper numbers are showing.

17 Now slowly flip up the right two tiles until you achieve the next state (figure 18).

18 Continue folding the right two tiles up until you have achieved the configuration of figure 19. This figure is refer-

red to as the **BOX** with **LID**. (See photo 1b.)

19 Placing your fingers inside the **BOX**, simultaneously pull outward on the center left and right hinges (figure 20).

20 Continue pulling at these points as shown until the next state is reached (figure 21). This position is referred as the **lazy L** (an L on its side).

21 The next move is a bit complicated the first time you do it, so proceed slowly. First separate the right bottom side of the **L** while holding closed the left lower corner of the **L**. The left center hinge should be lightly pushed in, and the top right hinge should be lightly pushed down. There is a color picture of this move partially performed on the first page of the color insert; it is titled "**Opening of L**". (See photo 1c.) When you are sure you have obtained the state shown in the photo, continue on by pushing in on the left center side hinge and pushing down on the right center top hinge. (figure 22.)

22 Now you must collapse this configuration by pushing down on the center of the upward right hinge while simultaneously pulling out the center of the left hinge.

23 Spin the puzzle so that you are looking at the back side (figure 24). Be sure the correct tiles are showing.

24 Fold the left front tile out and to the front (figure 25).

25 Continue folding this tile until the next state is reached (figure 26).

26 Turn the puzzle so that it lies in your hand as shown in figure 27.

27 Now fold down the right two tiles (figure 28).

28 Next, fold up the left three tiles (figure 29).

29 Flatten the puzzle until the **square solution** is reached.

30 CONGRATULATIONS!
YOU HAVE REACHED THE SOLUTION!

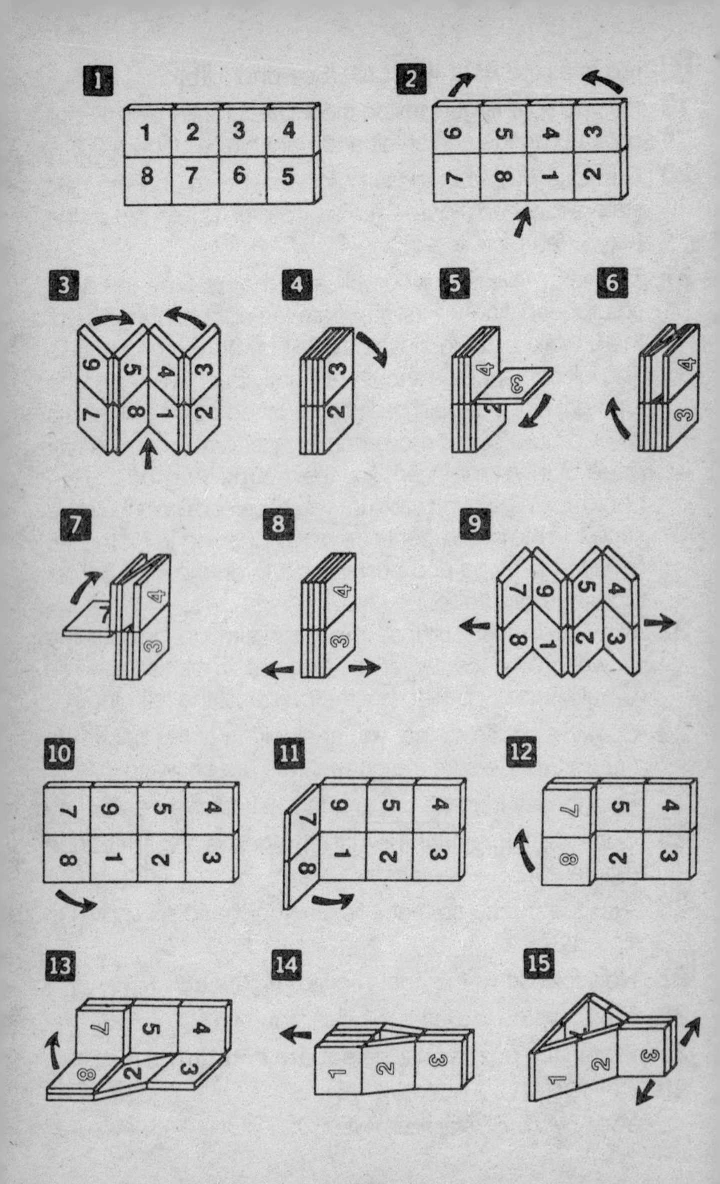
1
2
3
4
5
6
7
8
9
10
11
12
13
14
15

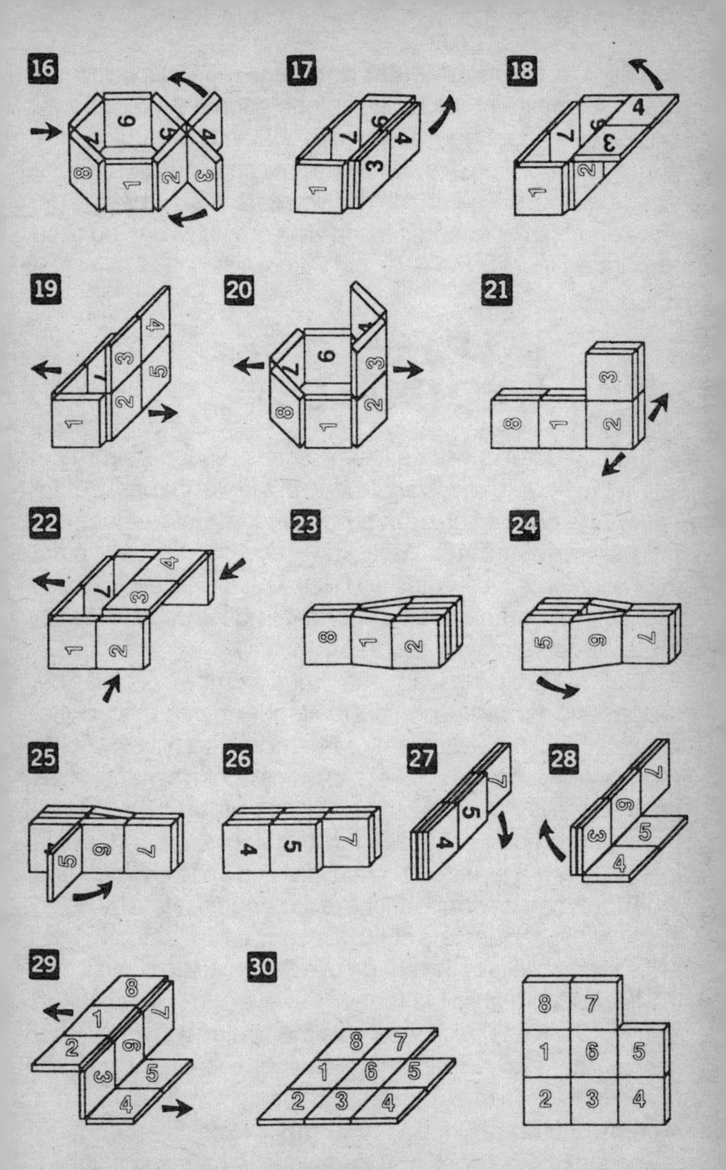
16
17
18
19
20
21
22
23
24
25
26
27
28
29
30

To return to the **rectangular solution**, you must follow the moves in reverse, checking the numbers again as you go. It will be easier to do if you go through the sequence using the numbering notation a few times before attempting to solve the puzzle without the numbers. The signature of E. Rubik, the patent pending marks, and the construction marks can be used to uniquely identify the **5**, **6**, and **7** front tiles.

7 HELP FOR THOSE WHO GET LOST

You have turned to this section because your puzzle is now lost in the three-dimensional world of Magic shapes. Before you will be able to solve the puzzle, you will have to return to the **rectangular state**. Work slowly and carefully, so as not to twist or force the hinges and tiles. The moves and shapes that are learned here are important in developing a better understanding of your puzzle.

Rubik's Magic has an enormous number of possible shapes and configurations, but remember that no matter what shape it is in, it is always only a few moves from many of the more familiar shapes you will see in this book. In many cases your puzzle is only a few moves from a **rectangular state**.

A common shape is the **Block**, two stacks of four tiles side by side. Many shapes can be transformed into the **Block** and the **Block** can be changed into a rectangular state by using many of the sequences of moves shown in this book.

If your puzzle is in the shape of a **Block**, try to open it into the flattest state possible (again, however, do not force it). If at first you are unable to reach a **rectangular state**, have patience. After playing with the puzzle for some time, you will gain a feel for the way it moves. Eventually you will reach a **rectangular state**.Once a **rectangular state** is reached, turn to Section 5 of the book and start working on the solution.